Contents

EARTH'S ROCK MAKES EARTH ROCK

Earth's Rock Makes Earth Rock

Do you know what the three different kinds of rock are? Check out page 7.

Do you know about Earth's layers? Check out page 8.

Do you know what sediments are? Find out on pages 11 and 12.

Read these words then check out pages 30 and 31.

core	crust	igneous rock	mantle
KAWR	KRUHST	IG-nee-uhs ROK	MAN-til

metamorphic rock
met-uh-MAWR-fik ROK

minerals
MIN-uh-ruhlz

sedimentary rock
sed-uh-MEN-tuh-ree ROK

strata
STRAH-tuh

Read this article to find out about rock. →

ROCK

Written by Collette Manners

What do you know about rock?
Have you heard the saying *hard as rock*?
People say hard as rock
because most rock is very hard.
Did you know that rock makes up
the outer layer, or **crust**, of Earth?
Earth's crust is between
6 and 40 kilometres thick.

Earth's Crust

The pebbles on this beach are pieces of rock. The sea has made the pebbles smooth.

Because rock makes up Earth's crust
you can often see rock.
You can see it in deserts,
on mountains, and in canyons.
Maybe you live near the beach.
If you do, it may be a pebbly beach.
Pebbles are small pieces of rock.
A layer of soil sometimes covers rock.
The layer of soil can be thick.
In many places, you would have to dig deep
to reach the rock.
Rock is under the sea, too.

What Rock Is Made Of

Do you know what rock is made of?
Rock is made of non-living things.
These things are called **minerals**.
Minerals are always solid.
There are a lot of different minerals.
Most rock is made of more than one kind of mineral.
Some rock is made of just one kind of mineral.

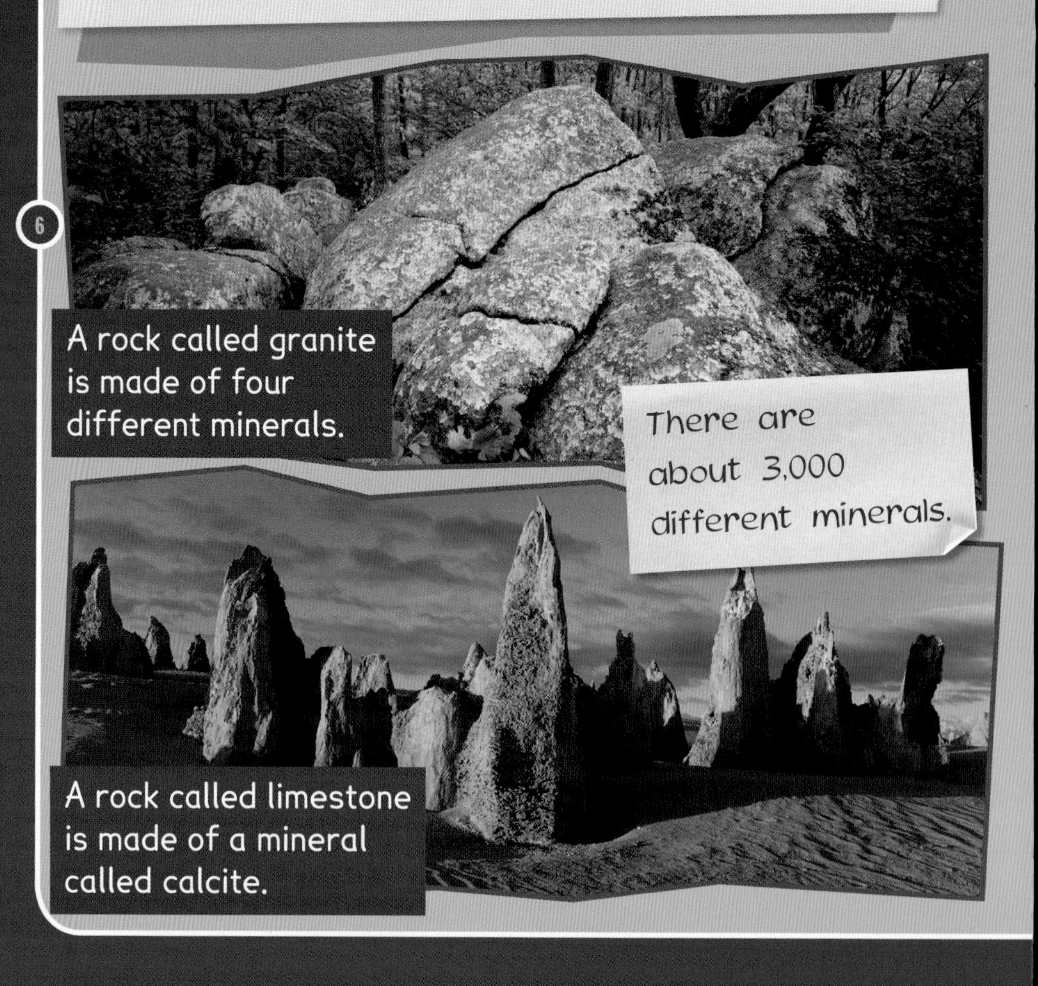

A rock called granite is made of four different minerals.

There are about 3,000 different minerals.

A rock called limestone is made of a mineral called calcite.

Kinds of Rock

There are a lot of different rocks.
All rocks belong to one of three kinds of rock.
They belong to a kind of rock
because of the way they form.
The three kinds of rock are **igneous rock**,
sedimentary rock, and **metamorphic rock**.

Uluru, in Australia, is made of sandstone. It is sedimentary rock.

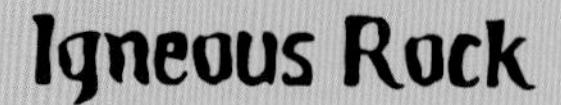

Igneous rock forms from hot rock
that is melted, or molten.
Molten rock comes from deep inside Earth.
It comes up to Earth's surface.
The molten rock is called magma.
Look at the diagram.
It shows you Earth's layers.
Can you find the **mantle**?
It is between Earth's crust and the **core**.

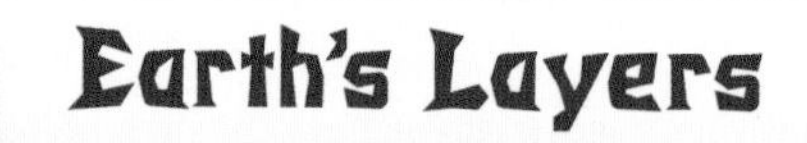

crust
inner core
outer core
mantle
magma

Earth's mantle
is 3,000 kilometres
deep.

Magma can be
as hot as
1,300 degrees Celsius.

The mantle is the layer that is magma.
Magma is very hot.
Magma mostly stays in the mantle.
But sometimes the magma
comes up through Earth's crust.
It comes out through a hole called a volcano.
The magma is now called lava.
When the lava cools down, it becomes solid.
Cool, solid lava is igneous rock.

This is lava
from a volcano in Hawaii.
It will form igneous rock.

There are two kinds of igneous rock.
One kind is on Earth's surface.
You can see this rock being formed.
The other kind is under Earth's surface.
You cannot see this rock being formed.
Look at the diagram.
It shows you the two kinds of igneous rock.

Kinds of Igneous Rock

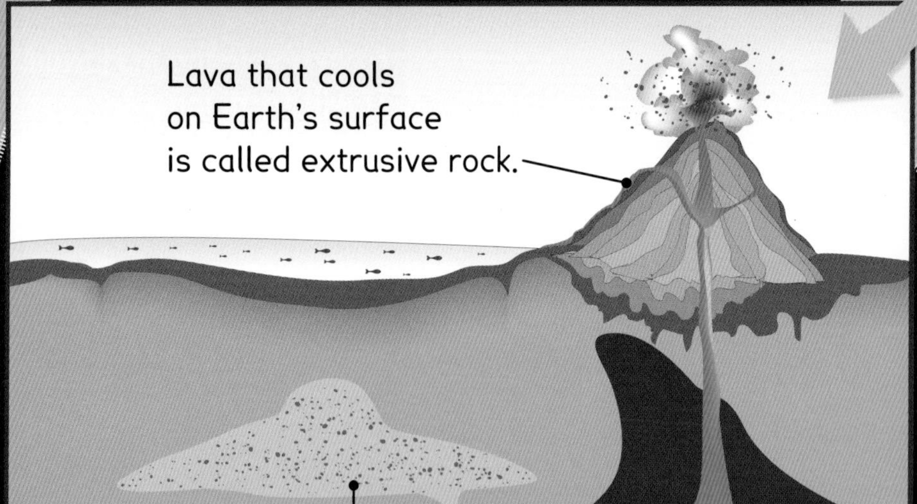

Sedimentary Rock

Sedimentary rock forms from bits of rock, sand, and clay.
It can form from bits of animal bones, too.
Bits of rock, sand, clay, and bones are called sediments.
They gather and make layers.
These layers are called **strata**.
Strata build up on top of one another.
They press down on the strata below them.
The strata become hard.
They form solid rock.

You can see the strata in this sandstone.

How Sedimentary Rock Forms

Over time, water, wind, and air break big rocks into small bits.

Water, wind, and ice carry the bits of rock down to the sea. They form layers of sediment.

When animals die, their shells and bones break down. The bits of shell and bone form layers of sediment.

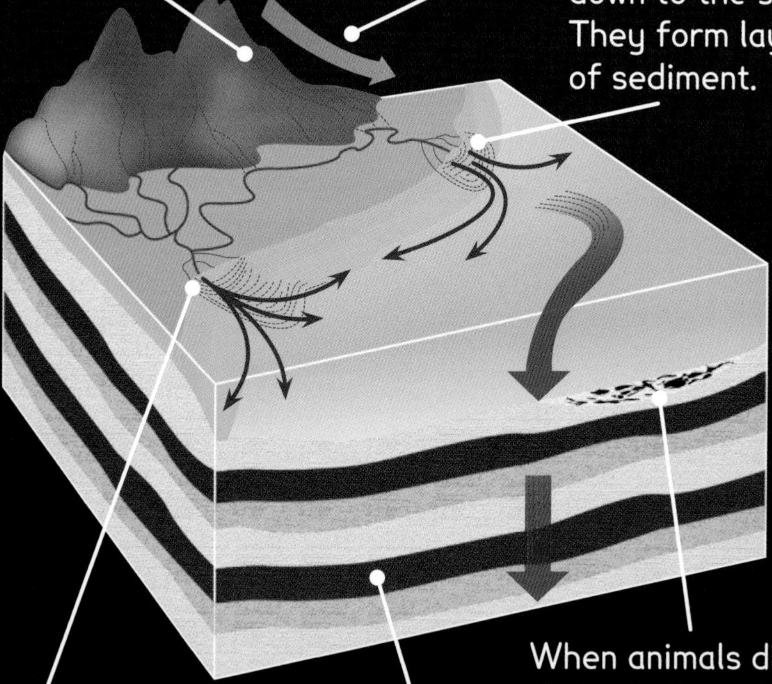

Chemicals come from minerals that dissolve in water. River water takes the chemicals to the sea. They form layers of sediment.

More and more layers build up. They press down on the layers below. This makes sedimentary rock.

Metamorphic Rock

Metamorphic rock forms when one kind of rock changes into another kind of rock.
Heat can change rock.
Pressure can change rock.
Look at the table.
It shows you where metamorphic rock forms.

How and Where Rock Forms

Rock	How It Forms	Where It Forms
Igneous	Hot lava cools.	On and under Earth's surface
Sedimentary	Sediments make layers that harden over time.	Under Earth's surface
Metamorphic	Heat and pressure cause change.	Deep under Earth's surface

Read on for a cave adventure. →

Rock 'n' Roll

Written by Collette Manners
Illustrated by Kelvin Hawley

What will we find in here? I'm looking for some granite.
Why don't you just look around?
We're in a big tube. Don't you think it's neat?
It's great, but why is it shaped like this?

Magma flowed through here. It cooled around the edges. Then the middle drained away. It left a hollow tube behind – a lava cave.
Cool!
Sure is.
KOOM

Was that...?

An earthquake! We need to get out of here, now!

That was scary.

Come on, Kata, get up! That may not be the end of it.

Where's the way out?
Don't panic.
Jon's right. We have to stay calm. It looks as if the earthquake has loosened some rock. It's blocked the way out.
Don't just stand there, Kata. Get out your rock hammer. QUICK!

OK, where do we start? Where shall we chip?
This is where we came in. So let's try above here. This rock is obsidian. It'll chip and break just like glass. Put your safety goggles on and let's get started.

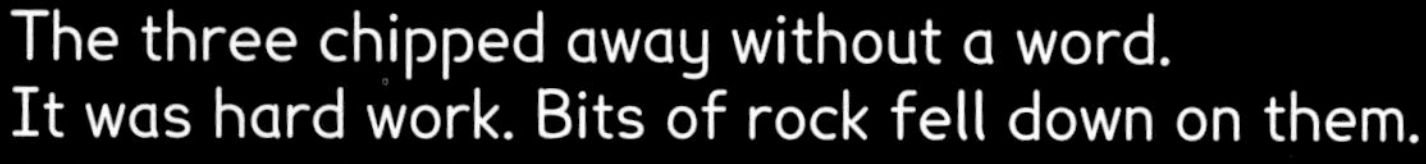
The three chipped away without a word. It was hard work. Bits of rock fell down on them.
TWUNK

PHEW!
HOORAY! Light!
We're breaking through!
We're nearly there.
KA-CHUK
They chipped away again and made the hole bigger.
KA-CHUK
You go first.

Turn the page to learn how one man survived being trapped under a rock. →

Tramper Cuts off Own Arm

Written by Collette Manners

Tramper Aron Ralston
cut off his own arm.
A boulder fell
on the 27-year-old man
while he was tramping
in a canyon.

Aron Ralston cut off
his right arm
below the elbow.

The boulder crushed his hand.
It pinned his arm.
He could not pull it out.
Aron was trapped for six days.
He ran out of food and water.
His arm started to rot.
He decided to cut it off.
He sawed through it with a knife.
Aron was free.
He was able to tramp out of the canyon.
Some trampers found him.
They called for help.

This is the place where Aron was trapped.

Read on to find out what the rock cycle is. →

The Rock Cycle

Written by Collette Manners

The rock cycle shows
how rocks form.
It shows how rocks change.
Rocks build up and wear down
all the time.
It takes a very, very long time
for rocks to form.
It takes a very, very long time
for rocks to change.

Look at the diagram.
It shows you the rock cycle.

Rain

Weather, such as rain,
slowly breaks down
the igneous rock.

Igneous rock

Lava cools
on Earth's surface.
It hardens.
It is now igneous rock.

Sedimentary rock

The rock breaks into small bits.
The bits form strata.
More strata build up.
The strata push down
on one another.
They slowly harden.
They form sedimentary rock.

Metamorphic rock

Over a long, long time,
sedimentary rock is buried.
It is pressed down. It heats.
It becomes metamorphic rock.

Magma

The metamorphic rock
becomes very hot.
It turns into magma.
It comes to Earth's surface.

Read on to find out about Earth's oldest rock. →

Multimedia Information

www.readingwinners.com.au

FAQS

Q What is the oldest rock on Earth?

A It is hard to say.
That is because Earth's rock changes.
It breaks down and builds up all the time.
However, rock over 3.5 billion years old
is found on all parts of Earth.

FAQS

Wind has worn away this sandstone rock over a long time.

~Rock Cycle Man Remembered~

James Hutton (1726-1797)

James Hutton was born in Scotland on June 3, 1726.
At school he loved science.
He wanted to know the history of Earth.
He thought he could learn this by studying rocks.
For years he studied.
He came up with a theory.
It was called the rock cycle.
It showed how Earth's rock
built up and broke down.
It showed how new rock
formed from old rock.
James Hutton died on March 26, 1797.
Since his death, the rock cycle
has led to a lot of new theories about Earth.

Turn the page to check what you have learned. →

1. How thick is Earth's crust?
2. Name the three kinds of rock.
3. What are layers of sediment called?
4. What are two things that change rock?
5. What shape is a lava cave?
6. What does the rock cycle show you?
7. How old is Earth's oldest rock?
8. Who came up with the rock cycle?

Turn to page 32 for clues. →

Learn More

Choose Your Topic

Choose one kind of rock in this book.

Research Your Topic

Find out more about this kind of rock.
Find out more about how it forms.
Find out where it forms.
Find out about a place
where there is a lot of your kind of rock.

Write Your Article

You may need to make notes first.
You may need to draw maps.
You may need to find photos.
You may need to draw diagrams.
Get your facts in order.
Use subheadings to help you do this.
Write a draft.
Check your spelling.
Check your punctuation.

Present Your Topic

Share your work
with other members
of your group.

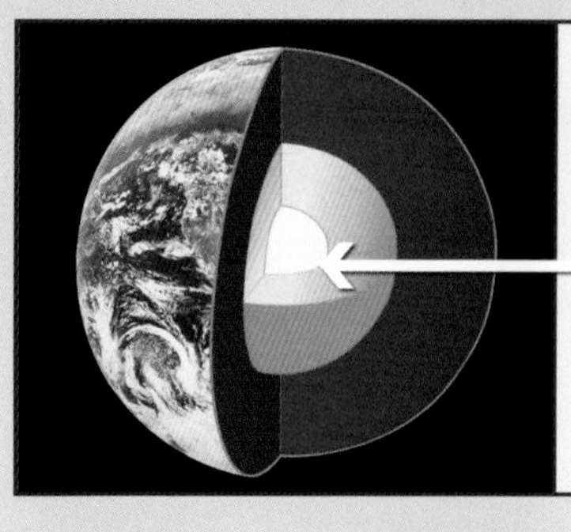

core – the middle of Earth

crust – the outer layer of Earth

igneous rock – a kind of rock that forms from molten rock that cools and hardens

mantle – the layer of Earth between the core and the crust

metamorphic rock – a kind of rock that forms when heat or pressure changes one kind of rock into another kind of rock

minerals – non-living solids found in rocks or in the ground

sedimentary rock – a kind of rock that forms when layers of sediment press down on one another

strata – layers of sediment

Index

Clues to the Quick 8 Quiz

1. Go to page 4.
2. Go to page 7.
3. Go to page 11.
4. Go to page 13.
5. Go to page 16.
6. Go to pages 24 and 25.
7. Go to page 26.
8. Go to page 27.